ELON MUSK

LEARN FROM
THE MASTERS

In the modern era, many people have become accustomed to taking part in engaging sessions that ensure they can come out of meetings with new ideas and the chance to genuinely improve and innovate within their business.

However, the reality of how difficult this is lost on most people – but not on Elon Musk. In the next pages, we're going to look at various quotes of his and try to decipher them for you.

As you read through you should be able to enjoy his unique take on various parts of society and business. Use these quotes to drive yourself forward to truly improving your current position.

Content

Quote #1

"IF SOMETHING IS IMPORTANT ENOUGH, EVEN IF THE ODDS ARE AGAINST YOU, YOU SHOULD DO IT."

What did Musk mean by this?

This quote is extra important to understanding both the man and his legacy. Musk is someone who takes particular pride in the companies he is involved in – not only are they companies with massive potential, but they have some very powerful and ethical solutions behind them, as well. His movement into lucrative markets such as Tesla energy ensures that Musk continues to offer something which could potentially change the entire world in time; even though the odds of being a success were very much against him.

The biggest meaning behind what Musk said here, though, comes from the fact that he offers this in just about every walk of business that he is involved in; everything he does is not only smart from a business point of view but it makes sense from an ethical standpoint, too.

What did he do about it?

He fought through adversity – if you take a look at what Musk can bring to the table in terms of helping you understand and improve your position as an ethical minded professional, it becomes much easier to understand what he done to make things work.

What Can Be Learned From This?

He ensured that companies, even if they were not the most profitable ideas to start with, could be nurtured and developed so that the positive effect of the system could easily be witnessed by everyone else.

In short, Musk spent a huge amount of his time working through the hardship of a failed idea until it would finally be a success. Most people would see this as madness but, really, it is what was needed at that moment in time to make sure that his companies could continue to improve and grow in the right direction as well as help to make the companies profitable.

Why are you not doing this today?

The truth is that most people don't do this, at all. They find that when the chips are down, so are they. The vast majority of people – supposed leaders – do not have the fortitude to follow through on the position they find themselves in. If you find yourself in this position and want to try and get out of it, then it makes sense to do so by sticking with the plan that you first came up with; nothing is likely to help you get through a tough period of work or something not working like having ethical motivation!

When you feel as if what you are doing is right and that it will bring an incredible range of assistance for people later on down the line, then you simply have to take that chance and try to take things on that extra level in terms of its overall importance. Instead of just going along with the development of an idea until you hit that first bump in the road, skip over those bumps and keep going; if your idea is good enough and is driven by smart enough reasons, you are much more likely to get it right.

Further Learning

http://www.entrepreneurs-journey.com/7524/why-some-people-succeed-against-all-odds/

http://www.inc.com/kelly-hoey/against-the-odds-entrepreneurs-globally-pursue-the-american-dream.html

http://www.ephemerajournal.org/contribution/fighting-against-all-odds-entrepreneurship-education-employability-training

Quote #2

"BEING AN ENTREPRENEUR IS LIKE EATING GLASS AND STARING INTO THE ABYSS OF DEATH."

What did Musk mean by this?

This quote is one that definitely stands out from the Musk collection – as far as putting together a bit of wording goes, this is one of his most enigmatic statements! The whole point of this quote, though, is the fact that so many entrepreneurs face just utter chaos every single day.

Even in your successful businesses it can become easy to lose track of the time and find yourself just dealing with a total overload – in many businesses, even just a couple of days can be enough to put things into a negative spin. The best part about this quote, though, is the fact that it so effectively breaks down the conceptions behind what Musk is trying to do – he's always engaged with activities which work bring most normal entrepreneurs to their knees. This is, however, the life that he chose.

Everything about this ensures that you will have no problems at all in delivering something truly exciting if you can get his meaning here – it might be a tough job, and it comes with lots of rather unpleasant "perks" but it's something that, if you want to be within this industry, that you will need to learn.

What did he do about it?

Well, he got on with it! Look at the massive empire that Musk has built today – near enough each of his decisions in terms of investment and business change have come from having the bravery and the fortitude to just get on with it. Musk has been involved near enough constantly with taking on difficult decisions and being "the bad guy" for many years now and it has, in truth, colored a lot of people's opinions of the man. However, what he has achieved throughout his business life is just sensational and is something that should be given immense credit.

What Can Be Learned From This?

The reason it deserves so much credit, though, is because he just dealt with the problems; he knew this could be a problem for him down the line, and that things would be tough, but instead of sulking he fought back and delivered a response which ensures he would have no issues of this kind in the future.

In fact, one of the most impressive parts of what Musk has done is the fact that he's battled through so many considerable challenges which greatly held back his chances of success.

Why are you not doing this today?

It could be quite simply as to why you don't engage with this kind of thinking today – there's no point for a lot of people. If you don't have the mental fortitude to always be dealing with What If and you never have the ability to move on from this, then you are not made out to be an entrepreneur.

As an entrepreneur you almost expects events to read the way they do when you are working with someone like Musk! Try and realize that being an entrepreneur today is all about being able to deal with the risks involved.

Further Learning

http://www.entrepreneur.com/article/238319

http://www.paggu.com/entrepreneurship/what-is-entrepreneurship/

http://onstartups.com/tabid/3339/bid/17741/The-11-Harsh-Realities-Of-Being-An-Entrepreneur.aspx

Quote #3

"OPTIMISM, PESSIMISM, F*CK THAT – WE'RE GOING TO MAKE IT HAPPEN."

What did Musk mean by this?

Well, it's a quote which is built around taking action – not enough people take action today and instead try to look at things in the glass half empty/full conundrum. The problem is that by breaking things down in such a mundane manner, you leave yourself open to having to feel like this one way or another.

The difference is that people like Musk are doers, and instead of worrying about the positive or the downside they just try it out – what's the worst that can happen? Optimism and/or pessimism are useless if you don't make the application of the process work in the first place!

This is what Musk is getting at – your expectations, good or bad, can be blown away in a matter of minutes if you're expected outcome is not the way that things turn out.

Instead of this it's always better to go in with the mentality of being able to get the job done instead of always concentrating on whether the response you get is *expected* to be a positive or a negative; it just takes that bit of time and patience to make this distinction count.

What did he do about it?

Well, he pushed through the adversity and just got on with delivering it! Look at what his companies have done – they don't make pompous statements or try to tell everyone they are about to unveil the greatest thing ever; they just do it. There's no inherent sign of anything going on with his companies and then BAM! – They release something new and exciting.

What Can Be Learned From This?

Rather than creating unneeded hype built on overly positive thinking, they create solutions that will be enjoyed by people as they create such a fantastic response to what is going to be needed to make it work.

Why are you not doing this today?

The simple truth is that most people are not suitable for this kind of thinking – they need to have expectation and build-up to ensure that something comes along in the right way for them. If you are not doing this today it's because you have likely never considered just taking action without building up a layer of hype and expectation before moving forward.

With this way of thinking, though, you can move beyond the conventional barriers that hold a lot of people back in their quest to make things a bit more comfortable for themselves – the best way to make this count for yourself, though, is to move through the process and clearly consider where you are at this moment in time in your process itself.

Are you overselling or underselling the potential of the project? Strip away your emotional attachment to it and the like, and consider the long-term benefits or struggles that it may face. If you can put yourself more in the position of being a buyer instead of the creator it soon becomes much easier to understand how to start effectively making these decisions to fit it all together and keep yourself comfortable within your business model.

Further Learning

http://www.forbes.com/sites/laurashin/2014/01/01/have-a-dream-heres-how-to-make-it-happen-in-2014/

www.lifewithoutpants.com/how-to-make-it-happen/

http://www.lifeoptimizer.org/2014/05/06/how-to-dream-big/

Quote #4

"AS GOD IS MY BLOODY WITNESS, I'M HELL-BENT ON MAKING IT WORK."

What did Musk mean by this?

Following on from the above quote, this one is an excellent little reminder of the fact that Musk works so incredibly hard — this cannot be argued against, not legitimately. His own performance is something that many people can find hard to escape, as he is so energetic and positive when it comes to making sure he has the right tools in place to make sure he can be a success.

His strongest tool, though, is his will and ambition — few people can match that and whilst it would be remiss to try and ask people to replicate that, it's the main driving force behind his incredible success in life. His ability to create dominant path into the industries he works within makes a massive difference to the overall landscape of modern technology, and it's all down to the fact he is so determined to be a success and to be someone who makes lasting change come to the fore.

If you need to understand why Musk is such a success, it's not down to good luck or anything like that — it's down to a work ethic that few can match.

What did he do about it?

Well, he worked! Look around for other quotes outside of this eBook from Musk and you will find it hard to locate a quote of him actively complaining or being native. However, he's also not some ray of shine optimist – he's realistic enough to know what he needs and what has to be put in place, but he's ambitious enough to not let anything stop him from achieving that. His determination and willingness to succeed drives him on through each day, and it's easy to see the massive range of positives that this can create for someone like Musk.

What Can Be Learned From This?

By enabling nearly everyone to get the help they need in buying into being more ambitious and taking slightly more risks with their aims and intentions, this quote can be a great way to help others see what Musk done with his own life.

He got the sleeves rolled up and started working harder than ever – the result? One of the most impressive companies on the market which is more than capable of delivering a truly staggering range of performance, ensuring that nobody has to fall behind this process and watch themselves struggle to make the grade again; it just takes hard work!

Why are you not doing this today?

It's easy not to do what you should be doing when it comes to this world; ambition is easy to find, determination is not. If you want to match your ambition to your determination then it has to be valued and looked after clearly; someone can find it harder to build up that plan and position in their mind to make sure it works out in their favor.

If you are not staying ambitious today then you will hit your glass ceiling sooner rather than later; take the time to understand this and drive yourself towards genuine prosperity.

Further Learning

http://www.amazon.co.uk/Hell-Bent-Success-A-M-P-Mills/dp/0956117902

elitedaily.com/money/entrepreneurship/ambition-important-success/

www.bbc.com/capital/story/20140805-ambition-born-or-bred

Quote #5

"FAILURE IS AN OPTION HERE. IF THINGS ARE NOT FAILING, YOU ARE NOT INNOVATING ENOUGH."

What did Musk mean by this?

The concept of failure is a very interesting one – to "normal" people the idea of failure is something they simply cannot accept. However, to others there is a genuine understanding that failure is something you need to go through if you wish to improve as a person – if you are not going through failure, how can you ever learn how to succeed? People who go through life without failure are going to find that they never improve and always stick to the same beliefs and ideals as they believe they are divine and cannot go wrong.

The problem is that this creates ego and arrogance and makes failure harder to accept – instead, others take the blame and the people just get on with it, believing they have already made it. A fear of failure is something that is prevalent in society and since it's seen as such a massive taboo, most people will avoid ever getting involved within the possibility of failing. What this does to you, though, is makes it nearly impossible to improve on your current position and reduces your chances of improving as a professional.

What did he do about it?

Well, he got smart! He failed!

Look at this career; many of his big startups – Tesla included – nearly went through the floor before they ever became profitable. He was willing to take the plunge and to risk failure in the hope that his idea would catch on and eventually become a success.

What Can Be Learned From This?

The whole ideology of what you get from being involved with this kind of system comes from the fact that failure is something that should always be seen as a learning curve instead of some horrible milestone that cannot be recovered from – failure is, for all intents and purposes, a good thing.

If you never fail in life then you can never learn – so this is what Elon done. He failed, he failed a lot. He made poor choices and learned from them – the result?

Some of the most popular businesses on the world, a genuine position as an industry leader and the small matter of upwards of $13bn in the bank; it's safe to say that the risks taken by Musk were calculated gambles that have paid off so much to create something which is genuine unique and interesting. If you need inspiration, look to failure.

Why are you not doing this today?

The simple reason is that most people are still hung up on failure being the end of all your days and your life – your progress and your reputation, in one, will be gone for good. Nobody will trust you again – who trusts the failure? This is the belief that most people are sadly left with when they decide to take anything on.

Instead of being so scared of failing, be excited about succeeding. A redemption story after failure that teaches you your strength and weaknesses is far more valuable to you than a startup business without any problems when starting.

Further Learning

http://www.innovationexcellence.com/blog/2010/05/06/innovation-failure-points-strangled-in-the-crib/

http://99u.com/articles/7072/why-success-always-starts-with-failure

http://www.success.com/article/why-failure-is-good-for-success

Quote #6

"IF SOMETHING'S IMPORTANT ENOUGH, YOU SHOULD TRY. EVEN IF YOU KNOW – THE PROBABLE OUTCOME IS FAILURE."

What did Musk mean by this?

That being afraid to try something is far worse than being afraid to fail. We spoke about failure in the above quote but this is about trying and being prepared to stick your nick out to see just how far you can take something – if something is important enough to you then there should be no hesitation and no worries about getting involved. If something that you see is going to be important enough to you then there should be no reason why you won't bother – even if you are sure it will fail.

Again, failure is not a bad thing – it teaches you more than you ever learn when you succeed. When you are a failure you learn why, and you get to see things from a new perspective. Musk wanted to raise awareness of this, and does by making sure that if something is going to be big enough and important to you and even to someone else then it's worth trying out even if you are never going to succeed.

It ensures that, without a doubt, you have no problems carrying things forward.

What did he do about it?

Well, Musk went about fixing this because he managed to find the right way forward and the correct path that ensures you can all just relax and have some fun instead of always being worried about if you are a success. Even if you fail but manage to raise a bit of wariness for the subject that you want to get involved with, was that not enough?

Also, remember that every success is not personal; there have been many professional failures on the part of Musk but they have created some kind of long-term legacy that will improve things.

What Can Be Learned From This?

This makes sure that time always has someone waiting to make that distinction, and is ready to actually make a stand. Leaving your morals at the door when trying to run a business is a poor decision and will, more often than not, lose your trust and your control over your clients.

If you want to avoid then then you absolutely have to understand what makes success different for us all. Some people find that success comes from being truly happy making a difference to their industry, whilst for others success is purely profit.

For Musk, though, it's so much better than this – it's both!

Why are you not doing this today?

It's quite common for people to make moral changes and not make a stand if they believe they will fail. After all it's better to succeed at something you don't believe in than fail at something you do believe in, right?

This is a genuine stance that many take and the fact is that it won't be any use to those who wish to master this if they are not willing to just be prepared to fail.

If you cannot take the right stance, then you should take the time to understand why it's so helpful.

Further Learning

http://www.ineedmotivation.com/blog/2008/04/why-you-need-to-fail-to-succeed/

http://www.scotthyoung.com/blog/2012/11/19/pursue-your-dream/

http://www.planetofsuccess.com/blog/2011/stop-worrying-about-failure/

Quote #7

"YOU WANT TO HAVE A FUTURE WHERE YOU'RE EXPECTING THINGS TO BE BETTER, NOT ONE WHERE YOU'RE EXPECTING THINGS TO BE WORSE."

What did Musk mean by this?

Many businesses today are built upon proverbial quicksand – they look great on paper but the reality is that they are sinking fast behind the scenes. Musk, though, wants to see more businesses take the route towards long-term prosperity even if it means they look less attractive to sponsors and big name PR firms in the short-term. It's better to have organic roots which are sustainable instead of a business which is good today but might not be here tomorrow.

Never take the time to start up a business on unsure footings or unstable ideologies – a business should always have a cast-iron plan as to how it's going to work otherwise it can suffer and struggle to be the best it can, regardless of any perceived quality – or lack – within the system.

If you need help in understanding what Musk may have meant about this, look at the difference between a debt-free business with plans to scale, and a business with debt that's reached its peak and won't be able to add on any more to its revenue streams or formats.

What did he do about it?

Well, he built businesses on long-term foundations; the short-term may not have worked out but the long-term has.

Now he is in charge of some genuine industry leaders which work out just the way they had intended, and he also runs and works with a whole host of different products and packages which are built upon the idea of creating a big change within their marketplace.

Should these predictions pay off then he will be the proud owner of businesses which are making massive profits and area also helping to shape the entire industry they are based within.

By doing this, you make sure you are left with an easy way to understand how a business should work.

What Can Be Learned From This?

This removes a lot of the challenges associated with business management as, so long as it's started and looked after in the right way, a business will have the ability to succeed in the long-term even if the beginning and the short-term quality is not quite up to the standards which were hoped for.

So long as the ideas are sound and the philosophy is fair, everything that you could hope to see from your business is going to be fully accessible.

Why are you not doing this today?

Well, the reasons are quite simple – fear and the inability to grasp this subject closely. Many people can find themselves dealing with solutions built around something totally different, and will usually look for the short-term boost hoping they can cover those long-term costs and expenses simply by being so impressive.

The main problem that lots of people will suffer from in this kind of world, though, is the inability to understand that only covering yourself leaves you vulnerable to suffering tomorrow. Avoid doing this as soon as you can by implementing a strategy which is built upon finances that you have instead of leaving yourself as king for the day, but with nothing in the future.

Further Learning

http://www.lifehack.org/articles/productivity/how-to-have-a-brighter-future.html

www.fastcompany.com/1465628/four-ways-create-better-future

http://www.marcandangel.com/2012/08/15/12-choices-your-future-self-will-thank-you-for/

www.ingramcontent.com/pod-product-compliance
Lightning Source LLC
Chambersburg PA
CBHW071836200526
45169CB00018B/1539